S0-DOO-424

AN OUTDOOR SCIENCE BOOK

GRASS AND GRASSHOPPERS

ROSE WYLER
PICTURES BY STEVEN JAMES PETRUCCIO

JULIAN MESSNER

 Published by Julian Messner, a division of Silver Burdett Press, Inc., Simon & Schuster, Inc. Prentice Hall Bldg., Englewood Cliffs, NJ 07632.

JULIAN MESSNER and colophon are trademarks of Simon & Schuster, Inc. Design by Malle N. Whitaker Manufactured in the United States of America.

(Lib. ed.) 10 9 8 7 6 5 4 3 2 1

(Pbk.) 10 9 8 7 6 5 4 3 2 1

Library of Congress Cataloging-in-Publication Data

Wyler, Rose.
Grass and grasshoppers / Rose Wyler ; pictures by Steven James Petruccio.
p. cm.
Summary: Describes the quantities of grass, how it grows, and the animal life found there.
1. Grassland fauna—Juvenile literature. 2. Grassland ecology—Juvenile literature. 3. Grasses—Juvenile literature.
[1. Grassland animals. 2. Grassland ecology. 3. Grasses. 4. Ecology.]
I. Petruccio, Steven, ill. II. Title.
QL115.W95 1990
591.5′2643—dc20
ISBN 0-671-66351-8 (pbk.) 89-12778
ISBN 0-671-66347-X (lib. bdg.) CIP
AC

The author and publisher thank Joseph Beitel and Michael Wong of the New York Botanical Garden, and Dr. Randall T. Schuh, of the American Museum of Natural History for their helpful suggestions.

Grass Roots and Shoots

Run through a green, grassy field.
Play and roll in the grass.
Hunt grasshoppers. Or sit down for a picnic.
The grass does not break. It springs back.

Cut the grass and it grows back,
taller and thicker than before.
Most plants die if they are trampled on,
or if they are cut. But not grass.
What makes grass so tough?
How does it stay alive?

Grass plants usually are hard to separate.
But if you dig up a clump and wash off the dirt,
you can take out one grass plant and see
all its roots, stems and leaves.
Just be sure you are allowed to dig up a clump.

Some grasses have two kinds of stems:
a green stem that stands tall and a brown stem
that creeps along the ground or below its surface.
Both kinds have roots that take in water
and hold the grass in the soil.

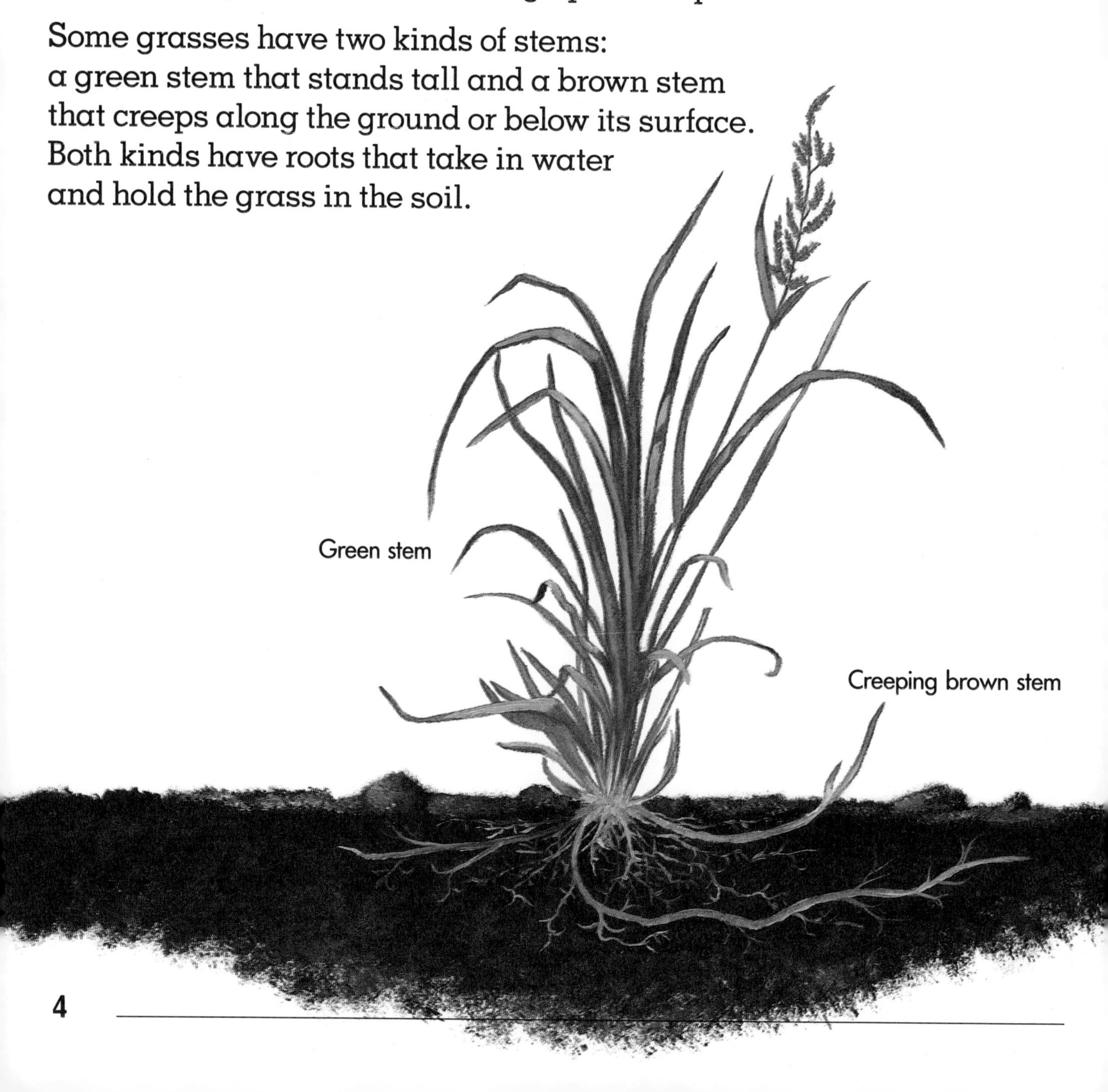

If you peel the leaves off a tall green stem, you will find little bumpy joints on it. Between the joints, the stem is hollow and is made of separate fibers.

◆◆◆◆

Cut off a piece of stem between two joints and see the hole at each end. Can you tear fibers from it? Can you blow through a cut stem? Try it.
Now plant the rest of the clump in a jar of wet dirt. Press down on the grass and watch it spring back. Its stem bends because it is made of separate fibers.

Grass is unusual in another way.
In most plants, the growing parts are the tips of the stems and leaves.
But in grass, the bottom parts of the leaves and stems grow too.
That's why grass can keep on growing if animals eat the top parts, or if a lawn is mowed.

Cut the clump of grass you planted down to the rim of the jar. Keep it in a sunny place and in a few days the grass will be above the rim again.

In some grasses, new shoots grow from
the base of the green stem.
in others they grow along a creeping stem.
Many grasses in lawns and fields have such stems.
In some kinds the stem creeps above ground;
in others it grows underground.
If you find grass with a creeping stem,
see how far it creeps.

Cows and other animals eat grass.
But how does grass get food?
It gets food the way all green plants do—
by making it from water and a gas in the air.

The food that grass makes is sugar.
Sounds impossible, doesn't it?
Yet every green blade makes a little sugar
all day long, all the while the sun is shining.

If grass is left in a dark place,
it loses its green color and cannot make food.
Without sunlight, its leaves and stems soon die.

What does grass do with the sugar it makes?
The sugar gets mixed with water in the leaves.
Sap forms, then flows into the roots and stems,
which use it as they grow.

Sap is also used to make flowers.
Each kind of grass has a different flower.
Here are some kinds that
bloom in lawns and fields.
They don't look much like flowers, do they?

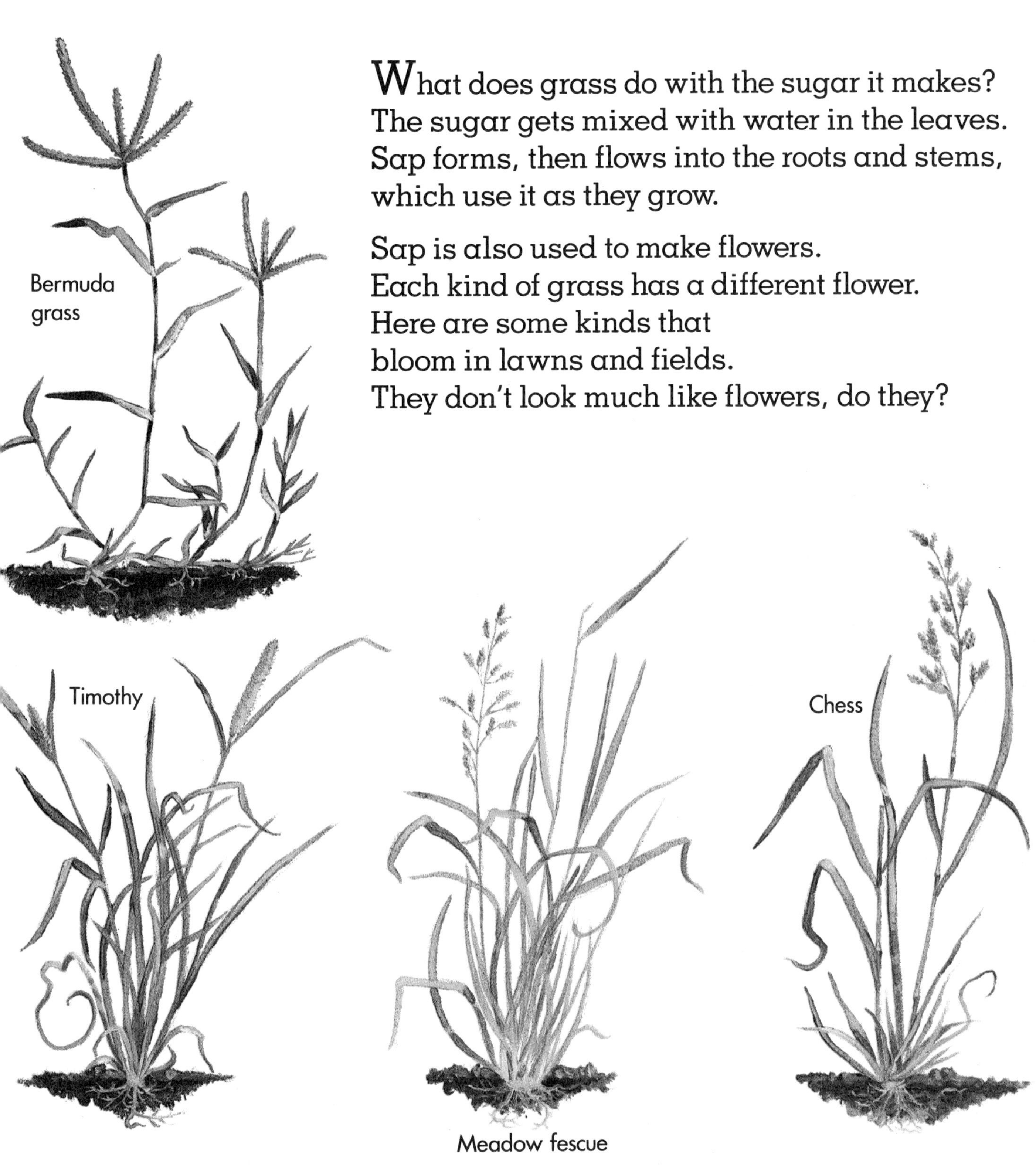

Grass flowers are green and have no petals. They are not showy like the flowers of buttercups, clover, dandelions, and daisies. But just like other flowers, grass flowers make seeds that start new plants.

If you find some dried-up grass flowers, you may be able to shake the seeds from them. Usually there are hundreds on just one plant.

Try this. Sprinkle grass seeds from outdoors or from a store on a wet sponge. Spell your name with the seeds if you wish. Then put the sponge in a dish with some water. Keep it in a warm, dark place. When sprouts appear, move the dish to a sunny place and watch the grass grow.

Outdoors, most grass seeds never sprout. Birds, mice and other animals eat them, and so the seeds are put to use.

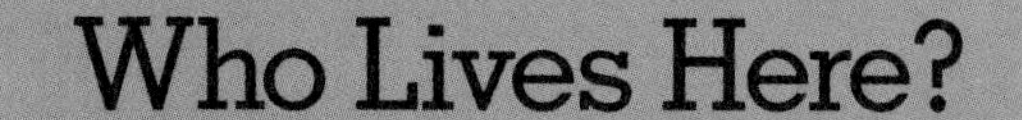

Who Lives Here?

If you look along the edge of a field
in late afternoon, you might see a rabbit.
Then again you might see only its tail.

Sometimes rabbits do not run away.
They just stay very still.
Since their coloring blends with the grass,
you may not see them.

Nearly all the creatures that live in grass blend with it. Their color protects them. Did you ever try to catch grasshoppers? There are lots of grasshoppers in summer fields but you only see them when they move. The way to catch them is to set up a trap.

For a trap, use a bottle. Bury it in the grass so that its top is level with the ground. A day or so later, a grasshopper may be inside.

After you trap a grasshopper, put it in a jar
with a plastic bag of ice cubes for a few hours.
Be sure the bag of ice cubes is closed.
The cold will not hurt the grasshopper.
It will quiet it so that
you can pick it up and examine it.

Notice the grasshopper's big pop eyes.
It sees in all directions with them.
It hears with ears set behind its big back legs.
The grasshopper pushes down with those legs,
then hops away.
Let it warm up and that's what it will do.

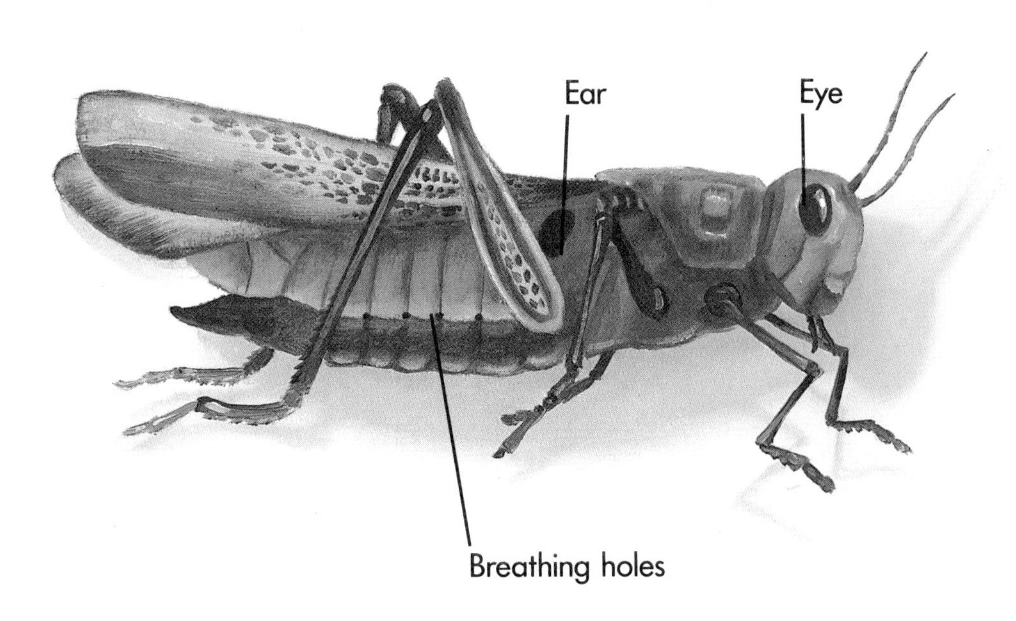

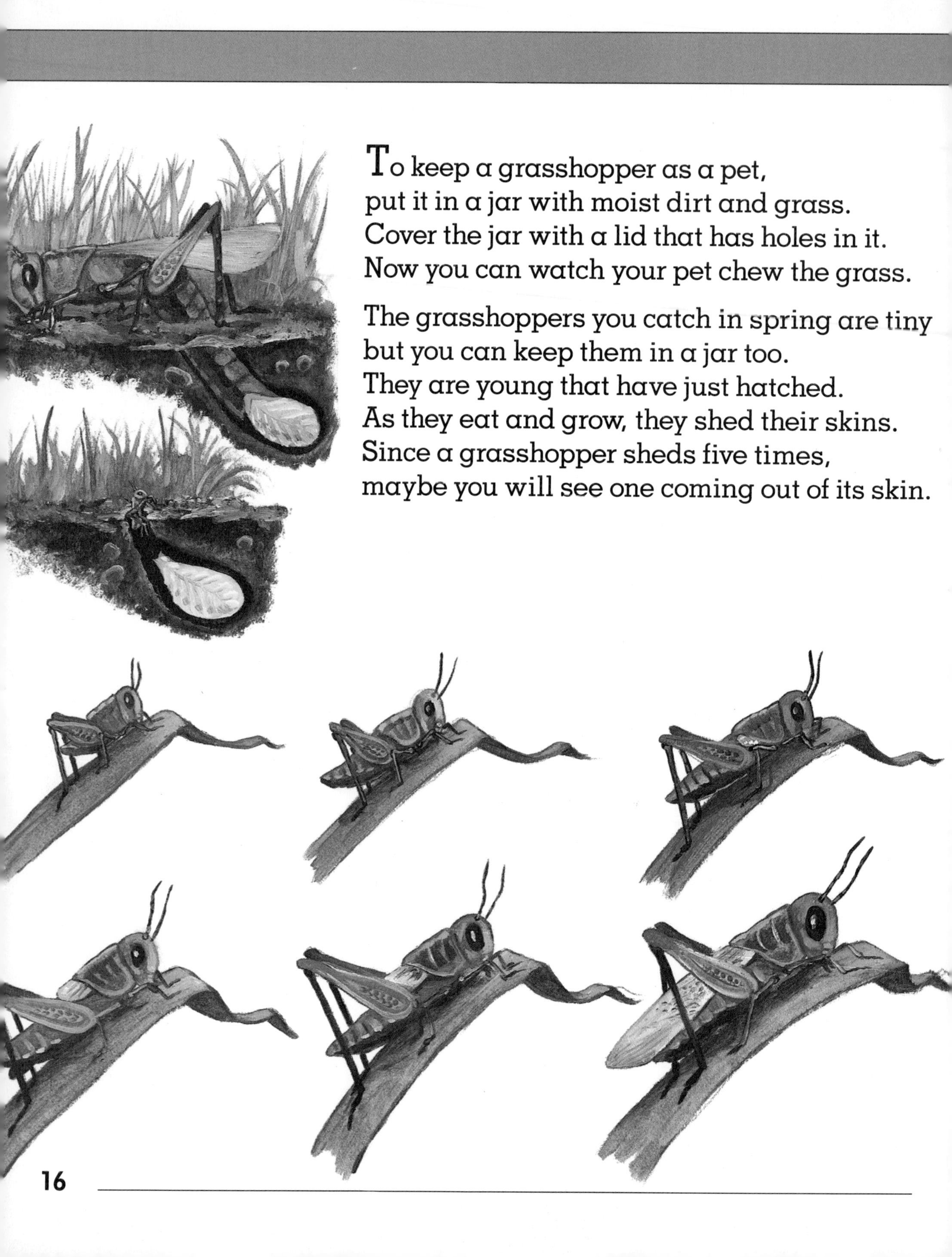

To keep a grasshopper as a pet,
put it in a jar with moist dirt and grass.
Cover the jar with a lid that has holes in it.
Now you can watch your pet chew the grass.

The grasshoppers you catch in spring are tiny
but you can keep them in a jar too.
They are young that have just hatched.
As they eat and grow, they shed their skins.
Since a grasshopper sheds five times,
maybe you will see one coming out of its skin.

Crickets look like grasshoppers but they are brown or black and have long feelers. Night is the time when crickets eat. They sing then too—that is, they make a cricking sound by rubbing their wings together. You can watch this with a flashlight. Move toward the cricket's sound, then stop when it stops, and move again when it starts. Move quietly and you can get very near the singing cricket.

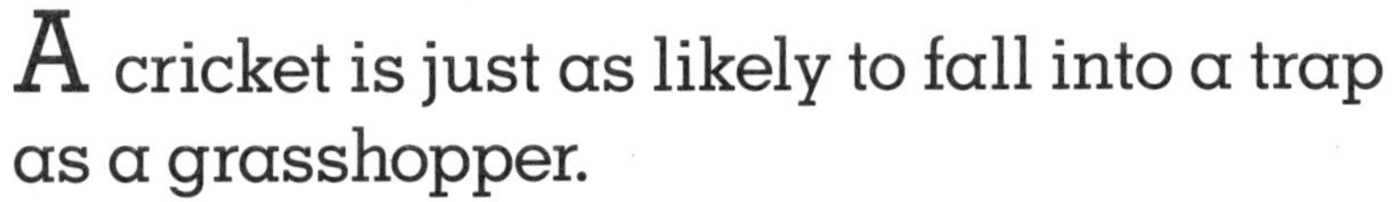

A cricket is just as likely to fall into a trap
as a grasshopper.
It makes a good pet too, if kept in a big jar
or bowl with sandy soil, twigs, water, and food.
For a cover, use a piece of wire screening.
Besides grass, the cricket will eat bits of
carrot, lettuce, cereal—even dog biscuit!

Will a pet cricket sing?
Yes, if it is a male; no, if it is a female.
A female has two prongs at the tip of her body.
She uses them when laying eggs in the fall.
If you keep a male and female, they may mate.
Then eggs will be laid right in the jar.

After cricket eggs hatch, the young shed their skin many times, as grasshoppers do. You often find these young in a trap in spring, along with other insects.

Did you ever see tiny bubbles on a grass stem? A small, young spittlebug makes the bubbles and covers itself with them while feeding. Move the bug to another stem and it quickly covers itself with bubbles again. Later, when fully grown, it will hop or fly away.

As soon as grass turns green in spring,
insects begin to eat it.
Caterpillars that look like little dragons
grow big and fat.
One kind will become the wood nymph butterfly.
One that is also found in gardens
will become the cabbage butterfly.

If milkweeds are in a field, you often find
striped caterpillars eating their leaves.
They will turn into lovely monarch butterflies.

Before a caterpillar becomes a butterfly,
it becomes a pupa with a tough skin around it.
The pupa changes inside the skin and wings form.
When the butterfly comes out,
it opens its new wings, then away it flies.

You can keep a caterpillar in a big jar with leaves of the plant you found it on. Put some holes in the jar lid. When the caterpillar stops eating, watch a tough case form around it. Weeks later a butterfly may come out.

If you find a woolly bear caterpillar,
do not expect it to become a butterfly.
It will turn into a pupa in a cocoon, and then
turn into a moth—a yellow Isabella.
How can you tell a moth from a butterfly?
A moth has feathery feelers; a butterfly has
feelers tipped with knobs.
When resting, a moth opens out its wings;
a butterfly holds its wings up straight.

You have seen little holes in a field with loose dirt around the rim, haven't you? Those holes are made by earthworms. Worms have no eyes but light bothers them. So they stay in their holes all day, eating bits of leaves they found during the night.

To coax a worm out of its hole, put a stake in the ground near the hole. Rub it with a stick. This upsets the worm and its head pops out. Give it a leaf and watch the worm pull it into its hole.

Wherever there is grass, there are ants, too. You often see a line of them, each one carrying a bit of food in its jaws. There is no fighting over the food, for ants live and work together in a group—a colony. Track them and you come to the colony nest. This may be a dirt hill or a tunnel under a stone with nooks for ants in different stages.

When ants hatch from eggs, they are larvae. The larvae grow until they become pupas. Then they change into adults.

Some ants sting, so be careful as you watch them. When a nest is disturbed, the adults get upset. They quickly move the eggs, larvae and pupas out of sight, to safety.

◆◆◆◆

Dig up a nest with the help of a grown-up. Put the ants and dirt in a glass jar with some crumbs. Cover the jar with cloth held in place by a rubber band and set it in a brown paper bag. In a few days you will see tunnels and ants at work in them.

When Winter Comes

In summer a lawn or field is a mini-jungle,
teeming with animals that hop, creep, and crawl.
But as days get colder, many animals disappear.
Grass stops growing and fall flowers—
asters and goldenrods—burst into bloom.
Milkweed now has pods full of silken seeds.
Monarch butterflies fly over them,
going south for the winter, as many birds do.

Grasshoppers and crickets can still find food,
yet grasshoppers do less hopping and
crickets do less chirping.
Fields become quiet. Winter is on its way.

As days get colder, they get shorter and there is less sunlight.
There is also less rain, so soil gets drier.
Grass stays green longer than most plants, but after a while it dries up too and stops making food.

What happens to animals that do not go south? Adult grasshoppers and crickets die, but the eggs they have laid survive through the winter. The pupas of some insects survive in cocoons. Worms just plug their holes and stay in them. For animals that eat seeds and other parts of plants, there is food for many months.

Some animals store seeds to use
when other food is scarce.
Squirrels sometimes dig in lawns, looking
for seeds they have buried.
Mice store food in their burrows,
but do not stay in them as some animals do.
Even when snow covers the ground, mice go out.
As they search for food, they leave behind
a trail of tiny footprints.

Tracks made by a rabbit are puzzling.
Two big footprints come before two small ones.
That's because a rabbit's big hind feet land
ahead of its forefeet when it hops.

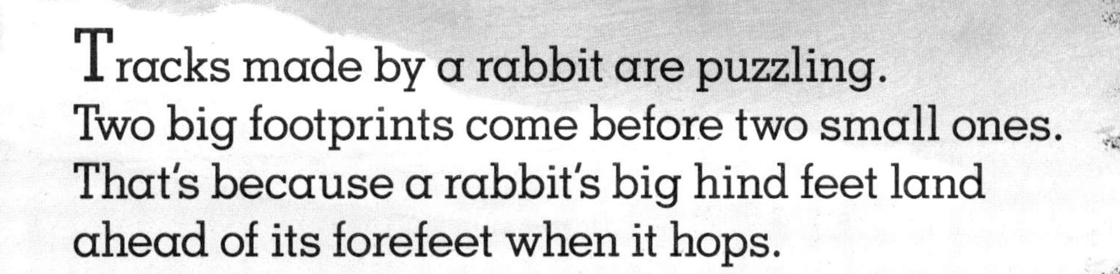

Pretend you are a rabbit. Get down on all fours. Then hop, bringing your feet down in front of your hands. Now you see how rabbit tracks are made.

When you find footprints, try to figure out
who made them and where they go.
Maybe they go to a burrow under a log or stone.

All winter long, grass seeds and creeping stems lie in the ground, ready to send up shoots. When spring rains wet the soil, grass shoots come up and leaves start making sugar. Soon grasshoppers and crickets hatch from eggs. Other insects come out of cocoons. Worms pop out and birds that went south return. Now animals hop, creep and crawl in green fields and life in the grass begins all over again.